I0755840

THE STRELTSY – RUSSIA'S FIRST STANDING FORCES 1550–1705

RICHARD L. SANDERS

The Streltsy-Russia's First Standing Forces

1550 – 1705

17th Century streltsy by Ludwig Scharf
(NewYork Public Library [NYPL])

Richard L.Sanders
Colonel, U.S. Army Reserve, Retired M.A., M.S.

Author:Richard"Rick"L.Sanders,Richmond,Virginia,USA

Printing and production: On Military Matters, Hopewell, NJ
ISBN: 978-0-965238-41-8

Originally published as "The Streltzi 1550-1705" as a three-part article in *Journal of the British Flat Figure Society*, No. 17 (Apr. 1990), No. 18 (July 1990) and No. 19 (Oct. 1990).

Edition: 2024

Front cover image "Soldier of the Strelitz Corps," probably Moscow 7th Regiment Stefan Feodorivich Yanov ca. 1674 (NYPL 960-1694 img 438642)

CONTENTS

DEDICATION

This book is dedicated to the late Heinz Tappert (Germany), whose interest in creating flat tin figures of streltsy led to my initial publication of this subject in1990; and to Colonel John Sloan, U.S.Army, Retired (Manassas,Virginia),who gave me the information, advice, and encouragement to pursue the topic of the streltsy.

ACKNOWLEDGEMENTS

My renewed thanks to Manfred Levec (Sindelfingen, Germany) for his editing and support with the research on the 2019 German publication that was key to this book. Of course, my very special thanks to the Swedish Army Museum (*Armémuseum*) for its permission to use their photographs of the streltsy trophy flags. My special thanks to Greg DiFranco (Smithtown, NY, USA) and Johan Lilliehöök (Åkersberga, Sweden) for the photos of their beautifully painted figures, and to Doctor Brigitte Grobe (Heinrichsen Zinnfiguren, Nuremberg) and Oleg Soktuto (Canada) for the photos of the old factory-painted Heinrichsen figures. Thanks to Mrs. Krista Wohlmann (Berlin) for the kind permission to display her and her late husband's figures. Also my special thanks to Markus Gärtner (Lampertheim, Germany) for decades of providing pictures of uniforms, many of which appear in this work. Last but not least to my wife, Ellen, for her patience and support in this project.

Streltsy types - watercolor by the author after contemporary drawings

A NEW KIND OF ARMED FORCE

The creation of Russia's first permanent standing infantry with firearms – the "streltsy" or "strelzi"(стрельцы,s-inglestreletsorstreletz,стрелец) – was among the military reforms of Tsar Ivan IV, the Terrible (1530-1584). The name streltsyis derived from the Russian *strela* (стрела), meaning "arrow" or "shooter," because these troops carried long firearms which distinguished them from other native Russian forces at that time.

The exact date of their creation is historically controversial, but it appears to have happened around 1550, when 3,000 men were selected from the already existing arquebus troops. Ivan IV probably based his new corps on the model of the Ottoman janissaries, and soon the Moscow Streltsy would assume a similar role in Russian society. Unlike earlier Russian fighting men, the streltsy were full-time soldiers whose units were not disbanded at the end of a campaign and not simply men called up to serve when needed. The streltsy corps trained regularly, and each regiment had its own artillery of six to eight cannons.

The streltsy troops were formed into six 500-man *statri* or districts, later called *prikazi* and then *polki*. They were stationed in Moscow in the Vorobieva Sloboda part of the city. During Ivan IV's reign, their number was raised to 7,000 men, of which 2,000 were mounted; and by the end of that tsar's reign, they numbered 12,000. In the summer of 1584, at the coronation of his successor, Tsar Fiodor Ionnovich, 20,000 streltsy were said to be present. Streltsy units were raised not just in Moscow but in other cities and towns. The day-to-day duty of the Moscow Streltsy was to guard the tsar's court and the cities and towns, to suppress internal revolts,and to protect the frontier until the entire army could be assembled. In peacetime, they were on permanent garrison duty, guarding the walls, towers, gates of the city, as well as government buildings. They also guarded the state saltpeterworks,convoys of money and prisoners,as well as escorting ambassadors. Besides the Moscow units, there were also *gorodovie* (town) streltsy. The Moscow Streltsy were without doubt in a privileged position so their pay, *dachi* (houses), and other perks were greater than that of the *gorodovikh* counterparts.

Streletz from the Savvino-Storozhevski Monastery (Zvenigorod), 1680s. (Palasios-Fernandez,R.,"Moskovskii Streltzi", p. 8)

In Moscow, the foot streltsy were on guard duty by weekly turns and were sent to strengthen the garrisons in othertowns.All larger cities had their own city streltsy – cities like Archangel, Astrakhan, Kazan, Novgorod, Pskov, and Smolensk. In the border towns, there were garrisons of twenty to one hundred streltsy which were mostly located on the northwest border in cities such as Pskov and Novgorod. Local streltsy units were also raised, e.g., in the early 17th century, the strategically important Novodevichiy Monastery had an independent streltsy garrison of 300 to 350 men.[1] There were fewer streltsy on the southern borders because the government had other troops such as the Cossacks in those regions. In 1682, Peter the Great sent some of the Moscow units to the southern frontiers

[1]Von Essen, *Muscovy's Soldiers*, p. 46.

to reduce their political influence in the capital.

The streltsy were recruited from among the freemen, and they promised to serve permanently. They were a caste of their own, often married, and their occupation became inheritable from the father to the son. To become a *streletz*, a newcomer needed a large number of sponsors from among the ranks. Peasants, serfs, and vagrants were not admitted. The streltsy came from the local people as a rule, but in Kazan, for example, 13 percent were arrivals from elsewhere. They had to be volunteers in good health and could shoot. The service, as in all military units of that time, was for life. However, it could be terminated at will by transferring the position to an heir. The heir of the deceased streltsy retained the house and land.

The commanders were from the *dvoriani* (court nobility) and from the *deti boyar* (literally, sons of the boyars, but really "hangers-on") classes, which were both types of hereditary service nobility. They received an annual salary of thirty to sixty rubles and a *pomestie* (estate given for service during the life of the person who served) of 300 to 500 *chertverts*.[2] The *sotniki*(centurians) received twelve to sixty rubles, and the *desnatyiki* (decurians) received ten rubles.

The simple streltsy's pay was four to seven rubles a year and twelve *chertverts* (ca. seventy bushels) each of rye and oats. The streltsy received that annual salary without having to pay any tax. The pay was so meager because Muscovy could not afford to pay them more and compensated with payment in kind, e.g., food. The annual state expenditures on the streltsy is estimated to have been 100,000 rubles.[3]

The streltsy lived in their own part of each town, physically separated from the rest of the population. Each man had his own house, yard, and garden. They also had land allocated to the unit for use of the members for farming. The size of the area varied according to rank and from town to town. The streltsy also enjoyed the privilege of trading without having to pay any tax or contributions. They engaged in small industry as well. Their commercial interests frequently brought them into conflict with the boyars and undermined their readiness for campaigns.

[2]A *chertvert* was an old Russian measurement equaling about six bushels.
[3]Koch,p.110.

Soldiers of various ranks from the Moscow regiments in parade uniforms

1. Polugolava (lieutenant colonel) of the 3rd Regt. Fedor Yashkin.
2. Standard-bearer of the 3rd Regt. with the sotnya standard.
3. Golava (colonel) of the 3rd Regt. Ivan Lopatin.
4. Colonel's bodyguard.
5. Selected strelets from the colonel's personal guards.
6. Strelets.
7. Standard-bearer with bratskoe (fifty-man unit) banner.
8. Uriadnik (senior lieutenant of fifty-man unit).
9. Sotnik (captain).
10. Drummer from the young streltsy.

Source:Palasios-Fernandez,R.,"MoskovskiiStreltzi,"pp.10-11.

ORGANIZATION

In its early years, all of the streltsy troops were directed first by the *Streletskaya izba*[4], then the *Streletskiprikaz*, mentioned for the first time on June 28,1682,when the Moscow Streltsy, having practically seized power in the capital, renamed themselves the *Nadvornaya pekhota* (Court Infantry) and they called their *prikaz* the *Pricaz nadvornoi pekhoti* (Court Infantry District). However, by December 17, the old designations were reestablished. In 1683, the *prikazi* were renamed "*polki*" (regiments) and the 100-man *sotni* became "*roti*" (companies).[5]

Streltsy, 17th century
Colorized, after a Viskovatov print

The streltsy were initially organized in six units of 500 men, called *prikazi* (also meaning district). The *prikaz* was further divided into units of 100 men called *sotnia* (or "hundreds"). The *sotnia* were divided into units of fifty men called *polisotnia*, which were finally divided into units of ten men, called *desyatoki*. In war, the city streltsy were designated to the various *polki* (regiments). The *polk* was commanded by a *golova* (pl. *golovoi* – headman or colonel), the *sotnia* by *sotniki* (centurians), the *polisotnia* by *pyatidesyatniki*, and the ten-man units by *desyatniki* (decurians).[6] When the originalsix *prikazi* were formed in the 16th century, the commanders and centurions were from boyar families. The first commanders of the streltsy regiments were Duma clerk Rzhevsky, Vasily Funikov-Pronchishchev, Fyodor Durasov, Grigory Zelobov-Pusheshnikov, Ivan Cheremisinov, and Yakov Bundov.[7]

Prikazi (regiments) were called by the names of the commander and had ordinal numbers beginning with the number one (1st). The lower the number, the more prestigious the unit's position. For example, for excellent service the 11th Regiment might be raised to be the 5th Regiment. In Moscow, the regiment designated as the 1st was the so-called *stremiannoiprikaz*, literally the "stirrup" court regiment, which by its strength exceeded the others by 1.5 to 2 times their manning. This unit's streltsy were partly or completely mounted, never sent out of Moscow to serve in the border towns, and always remained near the tsar's person.[8] The term "*stremiannoi*" (stirrup) came about because they were mounted and near the tsar's own stirrup. Based on the name, changes with turnover of commanders, and the shuffling of seniority of regiments under the honor system, one cannot anticipate continuity in the Moscow unit designations over time. For the streltsy outside of Moscow, there were often mounted units, but none could really be called cavalry; they were more analogous to dragoons, i.e., mounted infantry.

[4]An"*izba*" was a government office in Muscovite Russia.

[5]Palacios-Fernandes,p.8.

[6]Razin,vol.II,p.338.

[7]Lans,Olga,"RussianStreltsy–the Age of Emergence"

[8]Ibid.,p.8.

The first detachments of 3,000 men quickly proved their effectiveness. The streltsy regiments showed especially impressive results during the siege of Kazan in 1555.

1555 a Streletz at Kazan

During the Time of Troubles (*Smuta*, 1584-1613), whichlasted from the death of Ivan IV until Tsar Mikhail Fyodovich took the throne, Russia was in a state of civil war. Tsar Mikhail (1613-45) and his son Tsar Alexei (1645-76) reestablished order and modernized the army. They were aware of the value of a standing army, and without destroying the old army institutions, they established their own standing army composed of Russians and foreigners. They attracted Baltic German nobility to the officer corps and the state administration, and remodeled the army on the western European pattern. Many of the old external trappings remained, but the content was new – a basis upon which Peter the Great would build.
The modernized army numbered about 370,000 men, divided into four branches. The first branch, the streltsy and other old traditional troops, numbered about 130,000; of which, 49,000 were cavalry (mainly Cossacks) and 64,000 infantry.

Secondly, there were forcesof about 90,000 men (including50,000cavalryand 38,000 infantry) modelled exclusively on the western pattern. The third branch consisted of irregulars, mainly cavalry (160,000men), and the fourth branch of artillery (3,600 men). Even among the streltsy, western Europeaninfluences were evident. The *polk*, for example, was now organized like a regiment, and its leader, the *golava*, was renamed the *polkovnik*, like a colonel. The leaders of the *sotnia* (hundreds), called *sotniki*, became *kapitanii* (captains). Regarding unit strength, the information on the Russian army of the 16th century is very limited, and what little there is, comes mostly from foreign observers. Giles Fletcher said there were 5,000 Moscow streltsy, of whom 2,000 were mounted, while another observer said there were 10,000 streltsy in Moscow. By the end of the 16th century, they had increased to between 20,000 and 25,000 men.[9] At their peak in the mid-17th century, they numbered 50,000 cavalry and 45,000 infantry[10]. In 1663, there were 18,800 elite Moscow Streltsy and another 30,000 provincial streltsy, a three-fold increase from the mid-1630s.[11] By 1692, when rivalries between the streltsy and the boyars culminated in the streltsy revolt, Moscow alone contained thirty streltsy regiments totaling 30,135 men.[12]

---

[9]Sloan,"EvolutionoftheRussianArmy,"p.30.

[10]Koch,p.110.

[11]Davies,*Warfare,StateandSocietyontheBlackSeaSteppe*,pp.134-135,citingHrushevsky, "HistoryofUkraine- Rus," VII, pp. 216, 427, 432.

[12]Koch,p.109.

UNIFORMS AND EQUIPMENT

The clothing and basic equipment of the streltsy were supplied by the state. They were normally armed with long arquebuses, clumsy matchlocks (*samopali*) with straight stocks or later with muskets, as well as curved swords and *berdysh* poleaxes. Along with these, the state furnished them with one to two pounds of gunpowder, lead, and gunpowder flasks. The *berdysh* poleaxe had a handle which reached to about shoulder height,and the streltsy would rest his matchlock or musket in the crook of the axe's curved blade when firing or at the ready. The pole's butt sometimes was spiked to steady the axe when planted in the ground, and it could serve as a supplementary weapon in close combat. The *berdysh* had a sling attached to the pole so it could be slung over the shoulder, with the blade toward the bottom when not in use. The saber, for example, in 1684 had a Polish-style hilt, and some *gorodovie* streltsy (e.g., Savvino-Storozhevski Monastery in 1659) were armed with western European type sabers. Russian-manufactured matchlocks (*samopali*) were used by the streltsy only until the second half of the 17th century, and then they were gradually replaced with western European muskets which had larger calibers. The standard bearers and drummers did not carry long firearms but were armed with sabers or other edged weapons.[13]

Streltsy commander
probably from the 7th Moscow Regiment Stefan Yanoff (NYPL) (Artist: Jean-Baptiste Le Prince, 1734-1781)

Streltsy also carried finely-made daggers decorated with jewels and diamonds. In the 16th century, some streltsy were said to have carried "pikes" (*protazani*), but it appears they were more like halberds or spontoons. A contemporary lithograph from the 17th century shows a *strelets* with a halberd with a natural wood shaft, metal blades, and a large red tassel below the blade. Over the leftshoulder,the *strelets* worea bandolier withcartridges attached to the front and a small leatherpouch for flints at the right hip. A powder horn was also hung below the pouch or from a separate cord worn over the shoulder. The metal or leather scabbard with metal bindings was suspended on the left hip by means of two straps.

Officers normally were armed with a sword and some type of battle-axe, mace, orbattle pick, and they carried a walking stick or cane as a sign of rank – a custom which lasted for centuries in Russia. The senior streltsy commander was armed only with a saber. Other commanders were armed in addition to the saber with a well-decorated "*protazan*" (see page 15).

[13]Palacios-Fernandes,p.15.

Drummers and standard bearers are usually depicted as being armed with just a sword,but they probablyc arried daggers as well. The drums were rather small compared with those of western Europe, and they sometimes had rounded bottoms like kettledrums.

Regarding defensive personal equipment, there are sources that indicate standard bearers wore some cuirasses or body armor. Depictions, such as Viskovatov's plate No. 106, show streltsy wearing helmets in 1618 (image below right). One 17th century source described a march to the Kozhukhovskie maneuvers on September 23, 1694 as "*There went five streltsy regiments 1) thestirrup Sergueev, 2) Dementiev, 3) Zhukov, 4) Krivtsov, 5) Moksheev. All those five regiments made up 3,522 men. They were dressed in the old way (in East European dress – R.P.) in long 'polukaftanzhakh' (a type of kaftan), large baggy pants, with small helmets on heads, they wore their guns on shoulders and held in hand blunt pikes* [*kopya*]."[14]

Streltsy in 1618 (colorized Viskovatov plate 106)

Streltsy cavalry were certainly armed with sabers, and probably pistols as well, which would have been carried in saddle holsters. Much of the Russian cavalry of the 16th and 17th centuries was armed with lances and bows and arrows; however, the streltsy were probably distinguishable because they would have been equipped with firearms. The majorityof other equipment such as food, extra clothing, gunpowder kegs, tents, etc. were carried in baggage train. According to contemporary sources, their trains were much larger than those one would find with comparably sized west European armies.

Streltsy, being devout Russian Orthodox "old believers," wore their beards and hair rather long – a custom which led to controversy with Peter the Great when he began westernizing Russia.

Turning to the streltsy uniforms, there is very limited information about them in the latter half of the16th and first half of the 17th century, but there are some contemporary sources from the mid-1600s until the units' abolition in the early 18th century. Most notable among pictorial sources is the 1674 *Book of Travel Notes* by Erik Palmquist, who was captain of fortifications in Swedish service. In 1673-1674, he was part of a Swedish delegation that went to Russia. He was tasked with collecting information concerning Russia's roads and infrastructure, architecture, and technical equipment (including weapons).When he returned to Sweden, he compiled the album, including several drawings and maps, to report his findings.[15] A part of that work is shown on pages 22-24, and it forms the

[14] Ibid.,p.15.

[15]Palmquist,Erik, *NågraobservationerangåendeRyssland,sammanfattadeavErikPalmquistår1674* (Moscow: Lomonosov, 2012).

1600 Russian Streltsy 2nd Regiment, "Ivan Poltev"
(NYPL 960-1694 img 438649)

most definitive source on the Moscow streltsy as of 1674.

The contemporary written sources still in existence are mostly memoirs of foreigners who visited Russia at various times and a few remaining Russian documents with references to the ordering or supplies of streltsy uniforms. Unfortunately, the *Strelezkii Prikaz*'s archives were lost in a fire in the early 1700s.

The streltsy's issued clothing seems to have been semi-official from an early stage, with the *prikazi* and *polki* distinguished by the colors of their coats, coat linings and trim, hats, and boots. The long, heavy coat, called a kaftan, was ornamented on the front with button lacing which was silver or gold for the officers and in the distinctive color for the soldiers. The colored coat lining could be seen at the cuffs, which were sometimes sewn back and appeared like rounded cuff flaps. A 1588 work by D. Horsey mentioned the Moscow Streltsy in Ivan the Terrible's time as "*dressed very smoothly in velvet, colored silk and woolen dresses... thousands of Streltzi in red, yellow and light blue dresses, with brightening arms and arquebuses formed in lines by their chiefs.*"[16] The "colored" uniform was worn only on holidays. For daily service and on campaigns, they wore uniforms of black, gray, and brown shades. The officers' kaftans sometimes had a very long left sleeve, which could reach from below the fingertips, as far as to the knee! If it reached to the knee, a slit was provided on the inside of the elbow so the arm could be extended and the left band used. Such sleeves were traditional among the nobility and were quite common in their civilian attire. Some officers also had fur capes and lining to their kaftans. One contemporary source shows an officer with his coattails turned up with the corners tucked below his sash (see center figure page 5 for a reproduction). Inside the kaftan he wears apink smock and baggy dark green pants which could not be easily seen with the coat front closed. Whenever the baggy pants are shown in contemporary sources, they are worn tucked into the boots. The pants were probably in a distinctive color such as that of the hat or coat trim, that is if isolated, contemporary sources are indicative. The boots were probably made of felt or leather and were colored red, yellow, or green. At least one contemporary source shows brown boots.

Caps, also colored according to the *prikaz* or *polk*, were the standard headgear. It was pointed and was trimmed with sheepskin or more expensive fur for senior ranks. Officers had a small, flat, metal crown- shaped device in the front just protruding from the fur trim. Some streltsy wore rather modern-looking helmets, similar in shape to ones worn today in west European armies. They were smaller than the modern ones, fit closer to the head, and

[16]Palacios-Fernandes,p.11.

were made of curved plates bolted together (see the image on this page and on page 12).

Wide sashes were often worn over or inside the kaftan knotted in the front. They were of bright colored cloth, such as skyblue or yellow, but do not appear to have been uniform among the *polki*. Brown and black leather belts with metal buckles were also worn. Leather gloves with cuffs were also worn by some streltsy, especially by standard bearers. In the winter, all streltsy probably wore gloves or mittens.

Kemfer, another foreigner, who visited Moscow in 1661-1662, wrote, "*Their armament consisted of a matchlock, by which they used to salute, 'berdysh,' spiked in the ground before each, and had the form of half-moon, and a saber hung from one side. The kaftans were decorated well enough, one regiment was dressed in light green, another wore dark green cloth ornamented according to the Russian tradition with buttonhole lacing, which was gold and one chertvert* [ca. seven inches] *long.*" In light of this, it seems thatby the early 1660s, at least the Moscow regiments had uniforms distinguished by their color combinations.

1670 Streltsy in helmet
A contemporary print (Lars, Switki.ru)

Table No.1, gives uniform details for the fourteen Moscow streltsy *polki* as of 1674, according to W.W.Zwieguintsov.[17]

TABLE NUMBER 1

Regiment (number & name)	Kaftan	Buttonloops (troops)	Lining	Hat	Boots
1 Igor Lutokhin	red	raspberry	*	dark grey	yellow
2 Ivan Poltev	light grey	raspberry	raspberry	raspberry	yellow
3 Vassili Bukhvostov	light green	raspberry	raspberry	raspberry	yellow
4 Feodor Golovlinsky	cranberry	black	yellow	dark grey	yellow
5 Feodor Alexandrov	light red	dark red	light blue	dark grey	yellow
6 Nikofor Kolobov	yellow	dark raspberry	light green	dark grey	red
7 Stepan Ianov	light blue	black	cinnamon	raspberry	yellow
8 Timofei Poltev	orange	black	green	cherry	green
9 Petr Lopukhin	cherry	black	orange	cherry	yellow
10 Feodor Lopukhin	yellow-orange	raspberry	raspberry	raspberry	green
11 David Vorontzov	raspberry	black	cinnamon	cinnamon	yellow
12 Ivan Naramanski	cherry	black	light blue	raspberry	yellow
13 Lagovskin	dark red	black	green	green	yellow
14 Afanassi Levchin	light green	black	yellow	raspberry	yellow

* Zwieguintsov suggest that the lining for the first regiment may have been yellow since that colour is found in the regimental flag.

[17]W.W. Zwieguintsov, *L'Armée Russe, 1e Partie: 1700-1762* (Paris, 1967),p.3.

Senior officer, 3rd Viborgski Regt, 1696 with "protazan"

Little information is available about the dress of the streltsy cavalry. They supposedly wore red kaftans; however it is possible that they were also dressed just like the foot units. One source, Perle in May 1606, recorded the following:

"*There were formed in the lines Moscow Streltsy upto 1,000 men, in red cloth kaftans with white shoulderbeltsonthe-breast.Thosestreltsyhadlong matchlocks with red stocks; near them were 2,000 mounted streltsy, dressed in the same way as those unmounted, with bows and arrows on one side, and matchlocks attached to the saddle on the otherside.*"

Considering the number of streltsy cited here, it would be more than one *prikaz*, which would leadto the conclusion that at that time, the Moscow Streltsy may have just been wearing red and had relatively standard equipment and armament. The consistencyoftheredkaftansmayhavesimply been a result of an edict telling them all to procure that color coats and not necessarily that they were uniformly issued. Regardless, one can assumc that at lcast some mounted streltsy wore red kaftans and attached their firearms to their saddles.

Given the numbers of men noted as streltsy "cavalry" in thevarious sources, they obviously exceed the numbers of the Moscow "stirrup" regiment even with its augmented strength. So it is unclear whether the other mounted streltsy were from the fourteen various Moscow foot regiments or if they were raised, equipped, and uniformed separately. While sources agree that the mounted streltsy were more like dragoons than cavalry that would fight on horseback,one would wonder how difficultit would have been for them to be armed with matchlocks and *berdyshes* and be able to ride effectively.

(Top)
Streltsy cavalryman

(Bottom)
Dragoon of the Poselenski Regiment
Sovietski Voyn magazine, issue 3-91

THE STRELTSY BANNERS[18]

There were three types of streltsy banners: *Strelzkie "prikaz"* (regiment), "*sotennie*" (of hundreds), and "*bratskie*" (of fifty men). The regiment's banner was a well-decorated large colored cloth with pictures of religious themes and was carried out very seldom on special occasions, although given the number of such flags captured by Sweden during the Great Northern War, apparently those banners went to the field as well, in a couple cases in more than one flag, as shown in Appendix 2 on pages 47-52.

The "*sotennie*" banners, authorized for each *sotnia* (company) provided the permanent identification of the regiment. Their colors often coincided with the colors of the uniform, i.e., kaftan and cap colors. Finally, the "*bratskie*" banners, better said "flags" or guidons, were small square pieces of cloth decorated sometimes with a geometrical pattern, e.g., a cross (see page 8, figure 7).

The fourteen Moscow Streltsy regiments had flags with a common overall pattern during the late 1600s. It consisted of a central rectangular field, quartered by a straight cross, and surrounded by a border. Some flags had additional ornaments such as stars, crosses, and crescents. The flags of the Moscow regiments in 1674 are illustrated here and described in Table Number 2. Staffs were probably of natural wood,a metal spike or point was fixed at the top, and a colored cloth tassel was at the base of the pole covering. It was probablyof a like color. Some flags, such as those of the 1st and 10th regiments from Moscow were further adorned with cords and tassels attached below the staff head.

TABLE NUMBER 2

Regiment (number & name)	Field	Cross	Border	Corners	Stars & Wreaths
1 Igor Lutokhin	raspberry	white	yellow	---	---
2 Ivan Poltev	grey	raspberry	yellow	---	---
3 Vassili Bukhvostov	light green	raspberry	white	---	---
4 Feodor Golovlinsky	cranberry	yellow	white	yellow	---
5 Feodor Alexandrov	vermillion	yellow	white	light blue	---
6 Nikofor Kobolov	yellow	dark raspberry	light green	white	---
7 Stepan Ivanov	light blue	yellow	black	yellow	---
8 Timofei Poltev	orange	white	green	---	---
9 Petr Lopukhin	cherry	orange	white/yellow/ red	cherry	---
10 Feodor Lopukhin	orange	black	white/green	white	2 white, 2 raspberry
11 David Vorontzov	raspberry	black	white/light blue	yellow	---
12 Ivan Naramanski	black	white	yellow/cherry	---	cherry
13 Lagovskin	dark red	azure	white	green	white
14 Afanassi Levchin	light green	yellow	white	---	---

[18]W.W. Zwieguintsov, *Drapeauxet Etendards del'Armée Russe* (Paris, noyear),pp.2-3.

Streltsy with a sotnia banner, the regiment or town unidentified

The flags described here (Figs. A-F) and shown below were recorded by Zwieguintsov based upon records in the MilitaryMuseum in Stockholm. They are thought to be streltsy flags as well.

Flag A: Border in alternating chevrons of green, light yellow, and blue; in the corners, black squares. The central field is red with yellow crescents and stars. White cross with red ornaments. Violet staff sheath.

Flag B: White border with red flowers. Red central field with a gold cross. In the corners werered squares with gold stars. The flag was attached to the staff with blue laces.

Flag C: Green border; white central field with raspberry stars and black cross. The staff sheath was raspberry.

FlagD:Whiteandredborderwithsmalllight-yellowcross.Inthecornerswerelightbluesquares with red ornaments. The central field was light blue with a light-yellow cross with a small red cross in its center. The staff sheath was dark rose.

Flag E: The border was of alternating white and yellow chevrons. In the corners were dark blue squares with red flowers. The central field was dark blue with a red cross, and the staff sheathwas dark rose.

Flag F: The border was light yellow. The red central field had a white cross, and in the upperright quadrant were a white cross and stars. The staff sheath was raspberry.

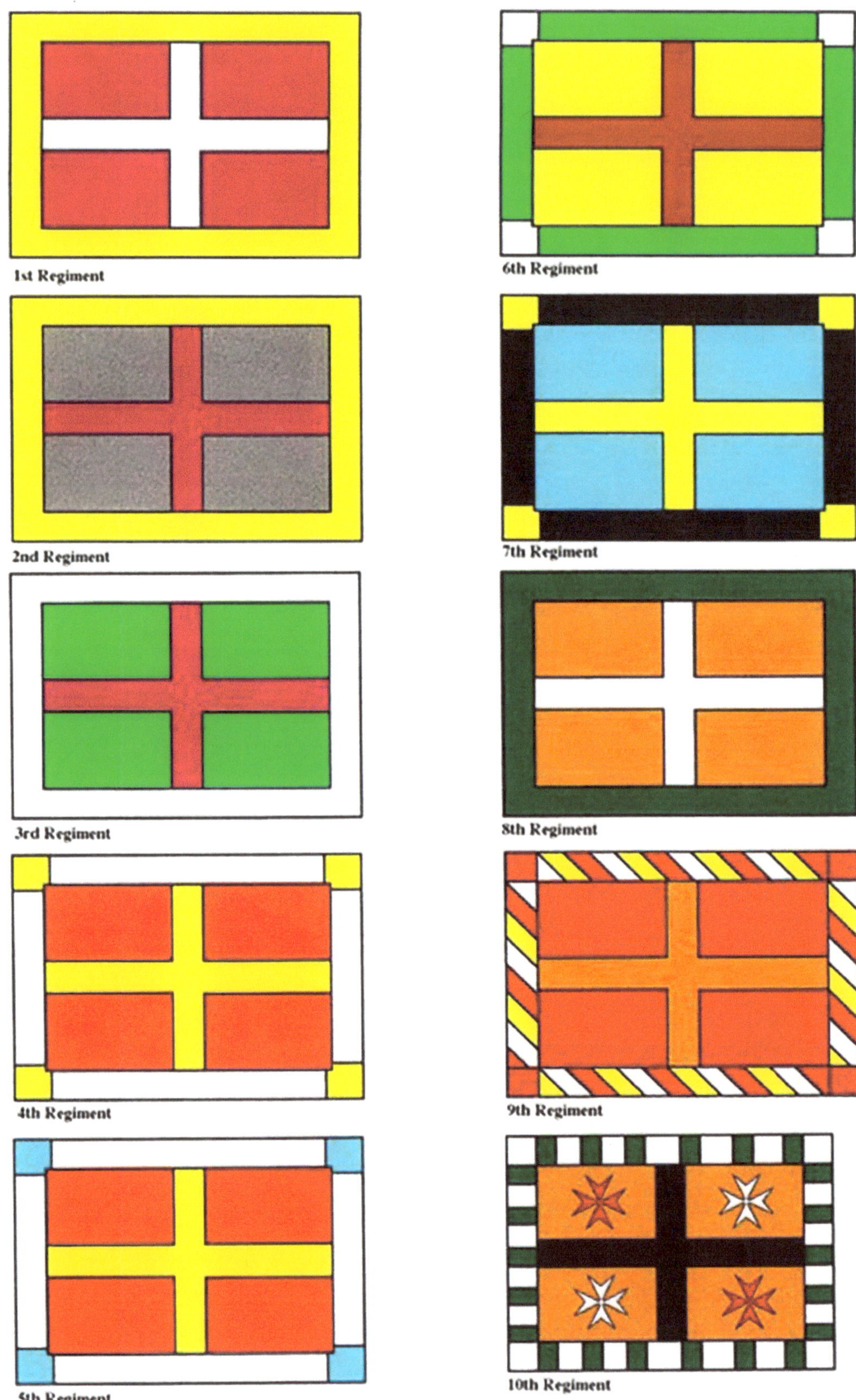

The flags of the Moscow Regiments 1-10

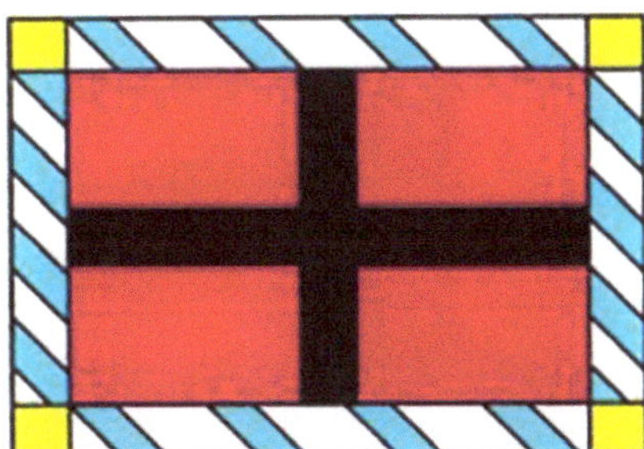
11th Regiment

12th Regiment

13th Regiment

14th Regiment

The flags of the Moscow Regiments 11 to 14

The following flags (A-F) are derived from Zwieguintsov drawings held by the Swedish Army Museum, Stockholm.

Flag A

Flag B

Flag C

Flag D

Flag E

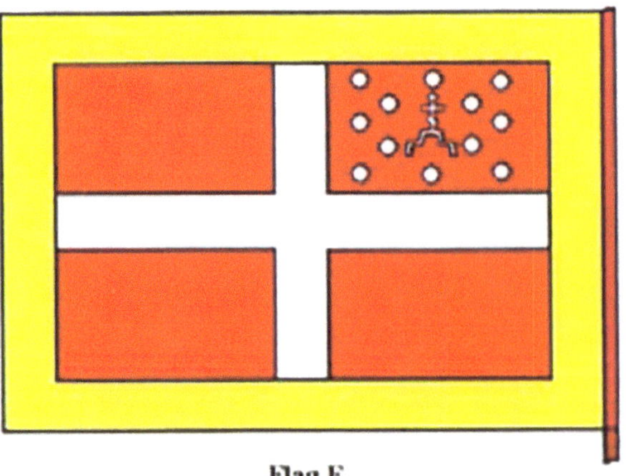

Flag F

The Moscow Streltsy Regiments and their Flags
As depicted by Erik Palmquist in his "Book of Travel Notes"of 1674

1st Regt. Lutochin
(NYPL 960-1694 img 439004

2nd Regt. Poltoff
(NYPL 960-1694 img 438665)

3rd Regt. Buchovastoff
(NYPL 960-1694 img 438666)

4th Regt. Golovlier
(NYPL 960-1694 img 438667)

5th Regt. Alexandrov
(NYPL 960-1694 img 438668)

6th Regt. Koloboff
(NYPL 960-1694 img 438669)

7th Regt. Janoff
(NYPL 960-1694 img 438670)

8th Regt. Poltioff
(NYPL 960-1694 img 438671)

9th Regt. Lopuchin
(NYPL 960-1694 img 438672)

10th Regt. Lupuchin
(NYPL 960-1694 img 438673)

11th Regt. Baranzioff
(NYPL 960-1694 img 438674)

12th Regt. Naramanskoy
(NYPL 960-1694 img 438675)

13th Regt. Lagoffskin
(NYPL 960-1694 img 438676)

14th Regt. Leoffskin
(NYPL 960-1694 img 438677)

On the following page, the following Moscow Streltsy regiments with their uniformsand banners are depicted based on Erich Palmquist's 1674 book, as presented in Palacios-Fernandes article:

1st *Stremyannoi* ("Stirrup") Regt., Igor Petrovich Lutokhin (1,500 men)
2nd Regt., Ivan Feodorovich Poltev (1,000 men)
3rd Regt.,Vasily Borisovich Bukhvostov (1,000 men)
4th Regt., Fedor Ivanovich Golovlinsky (800 men)
5th Regt., Fedor Vasilievich Alexandrov (800 men)
6th Regt., Nikifor Ivanovich Kolobov (900 men)
7th Regt., Staphan Feodorovich Yanov (1,000 men)
8th Regt., Timofey Feodorovich Poltev (800 men)
9th Regt., Peter Abramovich Lopukhin (1,200 men)
10th Regt., Feodor Abramovich Lopukhin (1,000 men)
11th Regt., David Grigorievich Vorontsov (600 men)
12th Regt., Ivan Ivanovich Naramansky (600 men)
13th Regt., Lagovskin (600 men)
14th Regt., Afasany Ivanovich Levshin (1,000 men)

Uniforms and *sotniya* standards of the Moscow regiments (*prikazi*) ca. 1674
Source: Palasios-Fernandez,R.,"Moskovskii Streltzi," p.14,illustration by E. Palmquist

TACTICS AND CAMPAIGNS

Little is known about thetactics employed bythe streltsy. Duringthe 16thcentury, Russian order of battle consisted of the center of peasant troops, dominated by the *gulay gorod* (a sort of"wagon castle"), and covered bythe streltsy and part of the artillery. About 1,000 horsemen were deployed in front of the center. Artillery was placed or entrenched on the left and right wings, flanked on each side by a *polk.* A rearguard which also served as the reserve was intended to be used as a surprise weapon in a battle formation which was inherently defensive.[19]

Gulai Gorod "Moving sledges into position"
from E.A Razin, *Istoriya Voennogo Iskusstva,* vol. III, p. 69.

"On campaigns and during sieges the streltsy were deployed in rows, usually behind the *gulay gorod*, or 'walking city.' The *gulay gorod* consisted of wood planks in a wood frame, assembled like a section of wall about 3 meters high and 3.5 meters long and pierced with a number of firing ports. The structure was mounted on top of a four-wheel or four-ski assembly so that it could be dragged by men. Along each vertical end of the *gulay gorod* were locking devices to allow then to be secured to adjacent sections."[20]

The use of such a mobile fortification made it possible to conduct more accurate fire and to be well-protected against enemy arrows and spears, as well as cavalry charges. The use of *gulay gorod* turned out to be very effective during the 1552 siege of Kazan, when the streltsy triumphantly captured the city, ending many years of wars with the Tatars.

---

[19]Koch,p.111.
[20]See"GulayGorod:BuildingtheRussianMobileFieldFortification," byCurtJohnsonand JohnSloanin*Gorgetand Sash*, vol .1, no. 1, pp. 21-25 for a detailed description and history of the *gulay gorod.*

The Battle of Dobrynichi, 1605
from E.A Razin,, *Istoriya Voennogo Iskusstva*, vol. III, p. 72.

Since the streltsy did not generally carry pikes – the standard infantry weapon of the 16th and 17th centuries – the *gulay gorod* served as a protection from cavalry charges, as well as giving cover from arrows and musket balls. A drawing from J. A. Razin's *The History of the Military Art* depicting the Battle of Dobrynichi (January 21, 1605), shows streltsy and artilleryin the field defending from behind a different kind of barricade against a cavalry charge. The barricades appear to have been made up of mounded dirt topped with straw. Three streltsy stand firing from behind each mound. Between the mounds are one or two wooden posts about 4 meters tall, held upright by guide ropes, and probably driven into the earth. These appear to have been used to anchor wheeled artillery pieces which appear between each set of mounds. This could have been used to reduce the recoil effect and hinder movement of the guns by enemies. A second row of streltsy stands about two meters behind those at the barricades and would replacethentoalternate in firing volleys, and they are backed up by another rank even further behind.[21]

During the campaigns against the Tartars and in the Livonian Wars, fortresses frequently had to be attacked or besieged. In the siege of Kazan by Ivan IV, a systematica nd protracted assault was needed, even though the tsar had 150 artillery pieces. During such sieges, all earthwork, mining, and sappingwas done by the soldiers and peasants, a job which west European Landsknechts and soldiers would seldom do themselves. The Russian forces simply dug their way through to the fortress walls, which in most cases in Russia were only wooden palisades anyway.

Several thousand musketeers participated in the Kazan and Polotsk campaigns of 1552 and 1563, although they may have not been streltsy as such, perhaps as arquebus-armed troops. In the summer of 1554, a Muscovite armyc onsisting of cavalry and streltsy was used in the conquest of theAstrakhan Khanate. The force went by river to the vicinity of Perevolok, wheretheVolgaand Don Rivers come the closest, on the border of that khanate.

[21]Razin,vol.III,p.73.

At the Battle of Molodi, which took place near the village by that name, about thirty miles from Moscow from July 28 to August 2, 1572, the streltsy and the *gulay gorod* were instrumental inthe defeat of the Crimean Tartars. The Russian army, outnumbered three to one, retreated, using the *gulay gorod* as a mobile fortress, surrounding it with hastily dug trenches. The main body concealed itself behind the walls of the *gulay gorod*, while outside, the remaining units covered the rear and flanks. At the foot of the hill on which the *gulay gorod* was deployed, 3,000 streltsy stood on the further side of a brook. The Tartar cavalry attacked and wiped out the streltsy at the brook, but the soldiers in the *gulay gorod* repulsed the attack with heavy fire from their cannon and muskets.

Battle of Molodi

The Tartars regrouped and rested for two days and resumed the assault on August 2 with infantry and cavalryundertheKhan's ownsons. Ignoringheavylosses, theTartarskept tryingto overturn the unstable walls of the *gulay gorod*, "*They reached their hands out to the fortress wall; manyTartars were killed, and countless numbers had their hands cut off.*" Late in the day, theRussians, who were running low on ammunition and food, executed a daring maneuver which decided the outcome of the battle. The commander and his forces secretly left the *gulay gorod*, leaving a small body behind, and made their way around behind the Tartars through a defile. At an agreed-on signal, the remaining forces in the *gulay gorod* and those at the Tartars' rear attacked and routed the surprised Tartars.[22]

In 1665, streltsy took part in the campaign in Podolia against the Crimean Tartars in Ukraine.[23] In the spring through autumn of 1677, during the First Russo-Turkish War (1676- 1681), 2,400 streltsy were among the forces in Chyhyryn who withstood a siege by the Ottomans.[24]

[22]Skrynnikov,pp.152-155.For a very thorough examination of this battle, see Muscovy's Soldiers by Michael F von Essen, pp.64-67.

[23]Davies,BrianL., *Warfare,State and Society on the Black Sea Steppe,1500-1700*,p.121.

[24]Ibid.,p.160.Chyhyryn is a city in Cherkasy Raion, Cherkasy Oblast,central Ukraine. From 1648 to 1669, the city served as the residence of the hetman of the Zaporizhian Host. After a forced relocation of the Ruthenian Orthodox metropolitan see from Kyiv in 1658,

Force structure changes were afoot in Russia after the wars with the Ottomans, trying to bring the military more in line with drill and tactics of foreign infantry formations. In 1681, a decision was made to restructure the streltsy-musketeers,ending their *prikaz* structure under *golovy* and puttingthem incompanies and regiments under captains and colonels, respectively. They were then ordered to"*study company formation frequently so that they will not forget it and will become entirely used to it.*"[25]

Chyhyryn under siege, 1674-78

it became a full-fledged capital of the Cossack Hetmanate.

[25] Ibid.,p.173,citing Fisher, Alan, *Azov in the Sixteenth and Seventeenth Centuries*, p.168.

THE DECLINE OF THE STRELTSY

For decades, the streltsy mixed in politics and, because of their commercial interests, they often sided with the people against the boyars and the tsar in the uprisings which marked the 17th century.

When Peter I (the Great) was a boy, he escaped death at the hands of the streltsy more than once. On May 15, 1682, with the connivance of the young tsarevich's stepsister Sophia, the streltsy stormed into the Kremlin to exterminate the family of Peter's mother, who was regent. The slaughter lasted four days, and Peter, aged ten, witnessed it. The fright from that incident left him with a permanent facial tic. A second such incident occurred in 1689, but Peter was able to flee with the aid of some of his soldiers.[26]

The streltsy were xenophobic, Orthodox "old believers" and opposed Peter's westernizing efforts which even included cutting off beards, which he saw as a sign of Russia's backwardness. Old believers,including many streltsy, considered Peter to be the Anti-Christ. In the summer of 1698, while Peter was touring western Europe, the streltsy started a revolt which increased in violence when a rumor started that the tsar was dead. Four streltsy regiments which were stationed along the southern frontier conferred with Sophia and marched on Moscow, de-

The Morning of the Streltsy Execution
Painting by Vasily Ivanovich Surikov (1848 – 1916)

termined not to let the tsar return. It was decided to put the Tsarevich Alexis on the throne, under the regency of Sophia, and to massacre the boyars as well as the foreigners living in Moscow. In June, forces loyal to Peter under General Boyar Shein and the Scottishgeneral Patrick Gordon put down the rebellion, executing many streltsy and holding many more in custody.

[26]Buehr, pp.137-138, 155, 156-157; *The Life and Times of Peter the Great* (Philadelphia & New York: Curtis, 1967), pp. 32-33.

Peter returned to Moscow in August and dealt with the streltsyin his own way. More than 900 of the streltsy lost their lives by beheading, hanging, or breaking on the wheel during the investigations of the cause of the rebellion. Evidence implicating Sophia, who had been lockedup in a nunnery during the revolt, and her faction was inconclusive, and Peter had the interrogations carried on for months in the hope of implicating her.

Peter the Great began the disbanding of the streltsy regiments in January 1699, immediately after the mass executions of the rebels. Some of the streltsy were sent to the county border towns. Some streltsy regiments were reorganized into combined-arms regiments with the preservation of a certain number of streltsy who had previously served in these regiments. There were regiments that were transferred in their entirety to the outskirts of the empire for garrison duty.

In June 1699, Peter disbanded Moscow's remaining streltsy regiments and dispersed the men and their families to distant parts of the country. The outbreak of the Northern War resulted in adjustments to the plans for the abolishing of the streltsy as a class. In 1705, the remainder of the streltsy units were abolished as a special corps, and the men were incorporated into the army which was reorganizing along western lines.

The streltsy fought in almost all the battles of the early eighteenth century. Nevertheless, by the 1720s, most of the units were abolished. But in some places in the hinterlands, the city streltsy served almost until the end of the century.

So the history of the streltsy ends in a less than glorious note. The tsar's lack of trust in the streltsy caused a decrease in the number of people willing to serve in these regiments. Peter did not want to recruit anyone and provide benefits. Young Russians preferred to serve in new modern units. By 1720, the streltsy troops had completely disappeared from the Russian Army.

CONCLUSION

The streltsy played an important role in Russia's history, providing excellent regiments in the mid-16th to mid-17th centuries. However, the Moscow regiments, like other palace guard forces, succumbed to corruption and began to interfere in political affairs. The Moscow regiments alienated one of the most talented and forceful rulers of their time, Tsar Peter the Great, when he was a child. As an adult, they opposed his efforts to westernize Russia and attempted tooverthrow him. It led to the execution of many of them and the internal exile of many others.

1.*Golova* (commander) of the 1st Regt. Egor Petrovich Lutokhin.
2.Standard-bearer with *sotnia* standard of the 3rd Regt.
3.Strelitz of the 6th Regt.
4.Streletz of the 13th Regt. in field *nosilnii* kaftan
5.Commander (of 500 or 100 men) 3rd Regt.
6.Streletz, 8th Regt.

Source: Palasios-Fernandez, R., "Moskovskii Streltzi," p. 13.

STRELTSY FIGURINES

"Flat" Figures As of the writing of this book, it is notable that miniatures oft he streltsy were produced at least 120 ago! The oldest figures identified in the research for this work were five miniatures produced as toys by the Ernst Heinrichsen Zinnfiguren (pewter figures) in Nurembergin1903.These are "flat" figures at the 28mm eye-level scale. Shown below, they are "streletz in firing position", "leader standing", "hornist", "streletz attacking striking" and "leader mounted" as representing them in 1630 and an extract from the listing. The molds for these figures still exist and it may be possible to obtain them again. Given that these are the oldest kind of figures identified, for continuity, the "flats" will be addressed first.

Streltsy by "Zinnfiguren ErnstHeinrichsen", 28mm, engraved in 1903

"Russische und andere Schlachten aus dem Mittelalter"

D	1156	1	russ	Strelitzen	1630	Sturm	zuschlagend	1903
D	1156	2	russ	Strelitzen	1630	Feuer	im Anschlag	1903
D	1157	1	russ	Strelitzen	1630		Anführer stehend	1903
D	1157	2	russ	Strelitzen	1630		Hornbläser	1903
D	1161	1	russ	Strelitzen	1630		Anführer zu Pferd	1903

After the flat figures of streltsy were produced by Erst Heinrichsen of Nuremberg in 1903 there appears to have been little or no production of streltsy until the 1950s or 1960s.

Later, as part of a wide series on the Northern War, the late Professor Walter Rössner (1925-1972) designed four 30mm streltsy figures and had the molds for three of them engraved. The fourth figure, a standard-bearer,was later engraved by Johan colored drawings for the four streltsy flat figures are shown here as well as the standard bearer produced and painted by Mr. Lilliehöök. The flag is based on a captured Russian streltsy flag now in the *Armémuseum* in Stockholm (see page 49). The other three figures can possibly be obtained from Erik Lindmark and Tenngjutaren.

A 28mm flat figure sold by Zinnfiguren Heinrichsenaspartofthe streltsy set Collection of Oleg Sokruto (Canada)

1600 Streltsy 30mm flat figure designed by Walter Rössner(†)
Standard bearer engraved, painted and photo by Johan Lilliehöök (Sweden)

The other source of flats suitable for use as streltsy produced in the 1950s or '60s by Harald Kebbel (Nuremberg). They were later available from SHA Zinnfiguren, Werner Fechner (Michelbach/Bilz, Germany). Herr Fechner cast several 16th/17th centuryRussian guards fromthe Harald Kebbel molds. He likewise has a number of civilian figures which nicely complement the streltsy. There is among them a mounted Boyar which works nicely as an officer for the streltsy (VH13a).

In 1983, at the Kulmbach *Zinnfigurenbörse* (Pewter Figure Fair) in then West Germany, this author happened to meet the flats producers Heinz and Erika Tappert (Duisburg) and mentioned the lack of figures available that depicted streltsy. Mr. Tappert suggested that this author design such figures and he would produce them.

Strelsty of the Moscow 14th Levchin, 8th Poltev, and 7th Ianov Regiments, ca. 1674
Figures in the St1 to St8 series produced by Heinz Tappert; designed and painted by the author.

The result was a series of eight 30mm flats based on the drawings in E. A Razin's *Istoriya Voennogo Iskusstva* produced in1984 as St1 through St8. Subsequently ,a far more skilled artist, Martin Block, designed the Tappert's streltsy for a total of over seventy types of streltsy and several types of civilian figure of this era. The figures designed by Mr. Block included a variety of marching, halted and firing troops plus officers, standard bearers and drummers. All the series is very historically accurate. Catalogue pages are shown in his work. Unfortunately, Mr. Tappert passed away in 2010 and it is unknown to the author where the forms have gone. The figures shown in this work may again be commercially available from a different vendor.

Heinz Tappert 30mm figures painted as 1674 Moscow 11th Regt. Voronzoff

Painted by the author

Heinz Tappert 30mm figures painted as 1674 Moscow 13th Regt. Lagofskin

Painted by the author.

At present, the most readily available streltsy flats today were produced by Mrs. Krista and the late Mr. Wolfgang Wohlmann in Berlin starting in the 1990s. They include 16th to 18th century streltsy and Russian Orthodox priests and are still being produced and marketed. The figures are depicted in this book are thanks to the kind permission of Mrs. Wohlmann. In April 2018, Ms. Krista Wohlmann handed over the molds and blank castings of the three series, Streltsy Attacking, Streltsy kneeling in prayer or surrendering, and Russian Orthodox Priests, to the private firm of Thomas Urbaniak, Ossenreyerstraße 3 A, 18439 Stralsund, Germany. The author thanks Mr. Thomas Urbaniak for his kind permission to reproduce the series in this publication. Furthermore, the editors of this book owe special thanks to Mrs. Krista Wohlmann, who, in additiontotheaboveinformation,alsoprovideduswiththecol-orillustrationsseries.

Series "Strelitzen im Angriff" (Streltsy Attacking)

Formerly WKW-Zinnfiguren, Berlin, now from Thomas Urbaniak, Stralsund

Painted by Wolfgang Wohlmann (†).

Series entitled "Streltsy kneeling in prayer or surrendering"
FormerlyfromWKW-Zinnfiguren,Berlin,nowwithThomasUrbaniak,Stralsund
Painted by Wolfgang Wohlmann (†).

While not streltsy cavalry per se, there are some 30mm flat figures of Russian mounted warriors from the same period that were produced in 1994 by Russian Historical Miniatures in St. Petersburg that might be converted to or certainly used with streltsy figures. One set is with the riders in trot and another set with them attacking. Catalog pages are at Appendix 3, pages 71-72.

Cavalry at the time of the streltsy by Russian Historical Miniatures, St. Petersburg,
Figures 476/94 to 480/94 designed by Alexander Mitelev (St.Petersburg,RF) and painted by the author.

"Full Round" Figures: Turning to "3D" or fully round figures specifically of streltsy, currently there are only a few fully sculptural figures on the market. Matchlock Miniatures (UK) makes some 15mm fully plastic "Wargaming" figure sets listed as "63X Muscovite Retainers," "64X Streltsi Musketeers," "65X Streltsi Axemen" with eight figures each, and "169X Streltsi Command" with six figures. A Moscow company, Zvezda, sells 46 unpainted, fully sculptured 1:72 scale plastic streltsy in a box. All are complete figures, except twelve of them, which are marching, have two arms separate that have to be glued on. Zvezda also produces acorresponding wooden fortress.

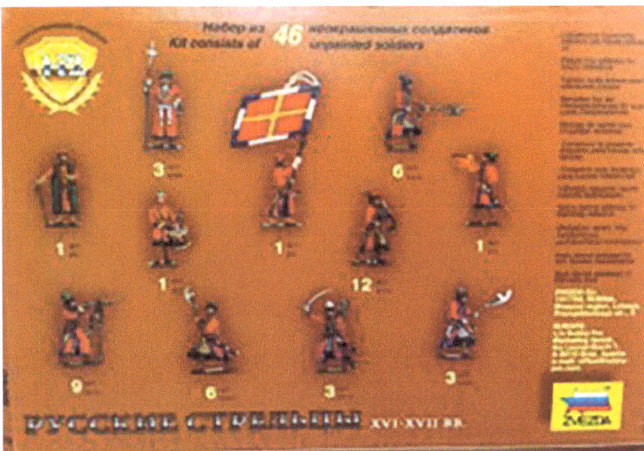

Zvezda's "Russian Strelets Infantry."

Figures painted by the author, displayed in front of Zvezda's assembled wooden fortress set.

Most recently, a British company called Highlands Miniatures, or 3D Printing Valley, produces a set of streltsy at the 28mm and 32mm scales. The so-called "Harbour Streltsy Unit (10)" models include a standard bearer, musician, an officer, and seven musketeers suitable for Warhammer gaming.The models are sold unpainted, printedwith a 3D printer in translucent resin and without bases. The figures are available from 3D Printing Valley Ltd, Karol Siemieniuk, Fibrocell Ltd, Unit7, Britannia Mill,Willow Street, Oldham, Lancashire, OL13QB, United Kingdom; Email: printingvalley@hotmail.com.

28mmHighlandMiniaturesHarbourStreltsyUnit(10)

Concerning larger scale "connoisseur" figures, the oldest one the author identified is a 54mm "Russian streltsy Infantry 1670" kit cast in a lead alloy made by J.D. Johnston, Leicester,England between 1969 and 1974, which might still be found online. Similarly, there was a 16-17 century "Boyar-head Streltsy" aiming a pistol, possibly produced in Germany with the identifier "Pol-008." Plus there was a streletz figure from the former Pelta Antaris company in Poland(next page).

Douglas Miniatures ~ Russian streltsy Infantry 1670
by J.D. Johnston, Leicester, England between 1969 and 1974.

"Boyar-head Streltsy,"
16-17 centuries, 54 mm,
"Pol-008."

Streletz of theTimeofy Poltevy Regiment, 1674, 54mm, Pelta Antaris,P oland

Painted by the author.

In 54 mm, from the EK Castings company of Yekaterinburg, Russia, there is a streletz with a shouldered musket (M200), a streltsy officer with spontoon (M201),and a senior streltsy officer halted (M202). These are one-piece castings and suitable for display without painting. They can be found online under "E K Castings" and are marketed through a company in France.

Left to right: 17th century streletz (M200), sotnik (M201), golava (M202).
54mm figures by E K Castings (Yekaterinburg, Russia)

Another purveyor of 54mm streltsy figures is Castle Miniature, located in Montreal,Canada.This firm produced at least two streltsy figures, one a streletz and the other an officer with a *protazan* (spontoon). Mr. Greg DiFranco (USA) converted the 54mm streletz figure below far left usingthe character below at the far right as his model. The resulting figure is shown below them.

Streltsy by Castle Miniatures, 54mm

Streltsy

Streltsy Officer
Figure converted and painted by Greg DiFranco (USA).

Further more,there is a75mm resin figure, as shown below,from an unidentified maker.This one was listed online for sale by someone in Germany and it could be that it is from a manufacturer there.

It is certain that there are more“3D” streltsy figures on the market today, but those described and shown here are the only ones the author was able to identify.

Streletz of 17th century, 75mm resin figure, maker unknown

BIBLIOGRAPHY

Barbour,PhilipL.,*Dimitry:CalledPretenderTsarandGreatPrinceofAllRussias,1605-1606* (Boston:HoughtonMifflin,1966).

Buehr,Wendy(ed.),*The HorizonHistoryofRussia*(NewYork:AmericanHeritage,1970).

Davies, Brian L., *Warfare, State and Society on the Black Sea Steppe, 1500-1700* (New York: Routledge, 2007).

Dmytryshyn,Basil,*AHistoryofRussia*(EnglewoodCliffs,NJ:Prentice-Hall,1977).

Essen, von, Michael Fredholm, *Muscovy's Soldiers: The Emergence of the Russian Army 1462-1689* (Warwick: Helion, 2018).

Graham,Stephen,*PetertheGreat* (NewYork:SimonandSchuster,1929).

Johnson,CurtandJohnSloan."GulayGorod:BuildingtheRussianMobileFieldFortress." *GorgetandSash*, vol.1, No.1, 1981,pp. 21-25.

Koch, H. W., *The Rise of Modern Warfare, 1618-1815* (Englewood Cliffs, NJ: Prentice-Hall 1981).

Lans, Olga, "Русские стрельцы: век появления" ("Russian Streltsy – the Age of Emergence"), Switki. ru website, Nov. 16, 2019.

Melegari,Vezio,*TheWorld'sGreatRegiments*(Milan:RizzoliEditore1969).

Mollo, Boris and John, *Uniforms of the Imperial Russian Army* (Poole, Dorset: Blandford Press, 1979).

Orlando,Enzo(ed).);BuzziGiancarlo;Johnson,Ben,*TheLifeandTimesofPetertheGreat* (Philadelphia&NewYork:Curtis, 1967)

Palasios-Fernandez, R., "Moskovskii Streltzii – Nepremennie Voiska Russkogo Gosudarstva XVII Veka" (Moscow Streltzi – Regular Troops of the Russian State in the 17thCentury), in *Tsoighauz (Arsenal) Military History Journal*, (Moscow, 1991), No. 1, pp. 8-15.

Razin, E.A., General Major, *Istoriya Voennogo Iskusstva* (The History of the Military Art), vols. II & I1I (Moscow: State Publishing House, Ministry of Defense, 1957, 1961).

Skrynnikov, Ruslan G., *Ivan the Terrible* (Gulf Breeze, Florida: Academic International Press, 1981).

Sloan, John. "Evolution of the Russian Army," *Gorget and Sash,* vol.1, No. 1, 1981, pp. 26-33;vol. 1, No. 2, 1981, pp. 10-20.

TheStateArmouryintheMoscowKremlin(Moscow:lzobrazitelnoyeIskusstvoPublishingHouse, 1967).

Zwieguintsov,Vladimir V.,*L'Arnée.Russe,1ePartie:1700-1762*(Paris,1967).

---*Drapeauxet EtendardsdeL'Armée -Russe*(Paris, noyear).

APPENDIX 1 – VISKOVATOV'S STRELTSY LITHOGRAPHS

From his *Historical Description of the Clothes and Weapons of Russian Troops*, 2nd edition published in 1899.

(http://warfare.tk/17/Viskovatov.htm)

Streltsy in1618
(lithograph106)

Kopiyitzik,1618 (spearman,in Moscow)
(lithograph107)

Musketeers of Moscow Streltsy regiments Lutokhina and Ivan Polteva in 1674.
(lithograph108)

Musketeers of Moscow Streltsy regiments Kolobov, Alexandrov,and Golovinskiy Bukhvostova in 1674.
(lithograph109)

Musketeers of Moscow Streltsy regiments TimothyPolteva,PeterLopuchin,Yakov,and Fyodor Lopuchin in 1674 (lithograph 110)

Musketeers of Moscow Streltsy regiments Lagovskina,Vorontsov,andNaramanskogoin 1674. (lithograph 111)

Standard-bearerZnamenshchikovandStreltsy Moscow Levshin Regiment in 1674. (lithograph112)

PrimarypersonsorofficersoftheMoscow Streltsy regiments in 1674. (lithograph113)

APPENDIX 2- FLAGS IN THE SWEDISH ARMY MUSEUM IN STOCKHOLM

Much of the information about the streltsy and other Russian flags of the late 17th and early 18th centuries is based on the collection in of battle trophies in the Swedish Army Museum (*Armémuseum*) in Stockholm. Many of those banners are from the Battle of Saladen (Saločiai or Salaty) which occurred in Lithuania on March 19, 1703 between Sweden and Russia, allied with Polish-Lithuanian troops. The battle ended with a Swedish victory, despite Swedish forces being numerically inferior to those of Russia. Most of the other flags attributable to streltsy units were captured by the Swedes at the Battle of Narva on November 20, 1700.

It should be noted that the flags on the following pages are more than 300 years old and in many cases were damaged during the campaigns and have faded over the centuries. It is also noteworthy that manyof these banners were exquisitely painted. The dimensions of the flags and the length of the flag's staff is provided if they were reflected in the Army Museum's digital listings (digitaltmuseum.se). The digital listing number for each flag shown is in the format of "AM.082nnn," and the flags can also be queried using the museum's "ST" number shown generally in the lower right corner of the image.

Probable streltsy flag taken at the Battle of Saladen or Saločiai on March 19, 1703. Dimensions: height 261 cm, width 360 cm, length of staff 295 cm. Armémuseum AM.082103

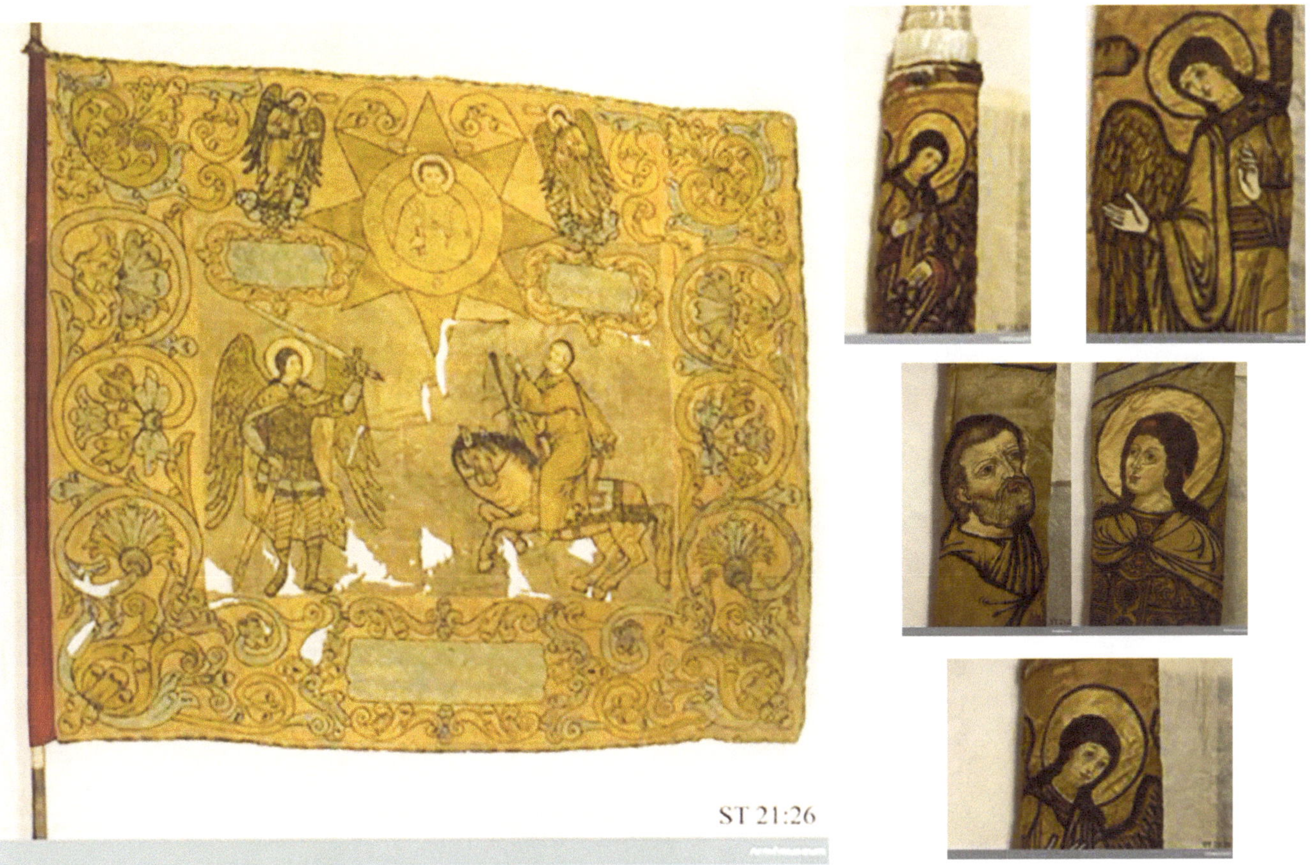

This banner was made for **Colonel Boris Fedorovich Dementiev's regiment of the Moscow Streltsy**. It was produced from September 1, 1692 to August 31, 1693. On the cloth there is a label with the following information: "Russian Province banner taken under King Carl XII's own high command at Narva on 20 Nov. 1700." It may have been taken at Saločiai/Saladen in 1703.

Dimensions: width 300 cm, height 265 cm, length of staff 310 cm.

Armémuseum: AM.082116

Streltsy Officer and streltsy, 17th Century

Those with red kaftans are probably from the 9th Regiment Lupochin; that in the grey kaftan from the 2nd Regiment Poltov.

(NYPL 960-1694 img 438638)

Above: **Regimental flag.** Museum's description reads "*Blue linen with red border of Chinese silk damask. Sheath of red broadcloth. The cloth is edged with 15 mm wide red fringe. Painted motif. The main motif depicts a Golgotha Cross with the torture tools. On either side were Emperor Constantine and his mother Helena. Above the cross is a semicircle with the judging Christ flanked by two seraphim. The flag was flown by a streltsy regiment from Novgorod under the command of Lieutenant Colonel Kristoffer Rikardovich Palmer. Taken at Narva on 20 November 1700.*"

This flag was taken to Stockholm in 1701. Produced on May 21, 1699. Dimensions: width 287.5 cm, height 276 cm, staff length 287 cm.

Armémuseum: AM.082123

Above right: "*Russian infantry flag of the* **Novgorod Streltsy Regiment** *under the command of Lieutenant Colonel Kristofer Rikardovich Palmer. Dated May 21, 7207 (1699). Taken in the battle of Narva on 20 November 1700. It is described in 'Truthful Story of the Arrival of the Russian Prisoners in Stockholm' printed in 1701 as:*

2 large provincial banners of pure sky blue and red around of damask in the middle of a large gilded cross. On one side of the cross stands Emperor Constantin, and on the other side his mother Empress Helena. Underneath the cross is a deaf skull. Above the cross is a painting as how one is supposed to paint God, and on each side an angel or cherubim painted and gilded. The length of the flag is 5 ells [cubits], the width 5 ells, the pole 8 ells long with a large elaborate gilded button."

Dimensions: width 279 cm, height 271 cm, staff length 327 cm.
Armémuseum: AM.082124

The banner was possibly made in 1670 for an **unidentified Moscow streltsy regiment**. It was captured by Lewenhaupt's army at Saladen on March19,1703.

Damask cloth. Dimensions: height 272cm, width 354cm, staff length 338cm.

Armémuseum:AM.082093

Saint Michael banner. These colors were both made for **Pyotr Ivanovich Borisov's regiment of Moscow Streltsy**. A sewn-on label on the cloth states that they were taken as trophies at the Battle of Saladen near Birzein Lithuania on March 19, 1703. They were produced on October 27, 1689.

Dimensions: width 311cm, height 273cm, staff length 360cm.

Armémuseum:AM.082114

Banner of an unidentified Moscow Streltsy unit. Taken at the Battle of Salaty (Saločiai/Saladen) in 1703. It was produced in 1681-82.

Dimensions: height 230cm, width 270cm, staff's length 310cm.

Armémuseum::AM.082091

The color was probably made between the years 1682-1689 for the lieutenant colonel's company in the **Stremyanni Regiment**, a mounted streltsy regiment from Moscow. It was captured by Levenhaupt's army at Saločiai/Saladen on March 19, 1703.

Dimensions: width 345cm, height 280cm, length of staff 370cm.

Armémuseum: AM.082094

This banner was taken at the Battle of at Britzen in Lithuania and the village of Schagarini, on March 19, 1703. According to probably erroneous information from the early 1800s, this Russian flag taken at Narva on November 20, 1700 under King Carl XII's Own High Command. There are three flags in the museum with almost exactly the same pattern. The inscription on this one is dated January 1, 1695 and indicates it was made for **Colonel Semen (Semyon) Matvevich Krovkov's regiment of the Moscow Streltsy**. The Swedish National Archives' list of the trophies taken in the Battle of Saločiai/Saladen mentions two banners of this appearance but it is most likely that all three originate from this battle.

Dimensions: height 284.5cm,width 317cm, length of staff 321.6cm.

Armémuseum:AM.082112.The other two flags, not shown are AM.082110 and AM.082111.

The Swedish Army Museum Infantry lists this flag as one for a streltsy regiment, possibly **Colonel Zacharii Vestov's regiment from Novgorod**, 1699.

Dimensions: width 199 cm, height 191 cm.

Armémuseum: AM.082204

Unidentified streltsy banner. In the museum's 1800 inventory, this banner is listed under the heading "*Trophies taken under Carl X's own high command in the Battle of Warsaw that lasted for 3 days in 1656.*"

Dimensions: height 258 cm, width 288 cm, length of staff 277 cm, length of secondary staff 312 cm. Note the decoration on staff, which appears on both sides.

Armémuseum: AM.082190

Both the flags shown above are listed in the museum's digital collection simply as a "**Russian streltsy regimental banner**". The one on the left (ST 21:101) has attribution to when or where it was captured. Its dimensions are height 230 cm, width 290 cm, staff's length 245 cm.

Armémuseum: AM.082191

The flag above, right is cataloged with the note "On the cloth is a printed label with information that the flag was taken at Saladen March 19, 1703. Its dimensions: height 252 cm, width 272 cm, staff length 320 cm; made of silk and linen, differ from those on the left.

Armémuseum: AM.082192

Unidentified Russian flag taken at the Battle of Birtzen (Saločiai/Saladen) on March 19, 1703…It was produced on December 18, 1689, according to an inscriptionon the canvas. Dimensions: width 300 cm, height 260 cm, staff's length 360 cm.

Armémuseum: AM.082099

Probable *sotnia* flag. According to a sewn-on piece of paper, this banner was taken in the Battle of Saladen on March 19, 1703. It was produced between 1600 and 1664 based on the type of cloth. Prior to 1664, the company colors for streltsy units were made of cotton, linen, etc. After that, they were made like the other silk banners. Dimensions: width 250 cm, height 250 cm, staff's length 370 cm.

Armémuseum:AM.082187

ST 21:98

Unidentified streltsy flag. There is no information about when the flag was captured. In older inventories it is stated as having been taken in Warsaw in 1656, but its pattern shows great similarities with two colors taken in Narva in 1700 (below) (ST 21:33 & ST 21:34). Dimensions: width 203 cm, height 191 cm, staff's length 224 cm.

Armémuseum: AM.082188

Saint John the Warrior (left) and Archangel Michael (right) flags. According to the Swedish Army Museum, a banner of the type was used by streltsy units. Made with silk taffeta cloth. The inscription translates to "God Father… The Holy Great Martyr John the Warrior."

It is unknown for which unit or units the flags were made. The type with a four-part cloth and border is common for Moscow streltsy throughout the 17th century. Saint John the Warrior was particularly revered in the region of Little Russia, which possibly says something about its provenance. It was captured at the Battle of Narva on November 20, 1700 and included in the 1703 Armory inventory. This flag was produced between 1650 -1700.

Dimensions: width 210 cm, height 200 cm, length of staff 335 cm.

Armémuseum: AM.082121 (left) and AM.082122 (right)

Unidentified streltsy flag. According to the label sewn on the cloth, the banner was taken in the Battle of Saločiai / Saladen near Birze in Lithuania on March 19, 1703. Dimensions: width 240cm, height 225cm, staff's length 345cm.

Armémuseum: AM.082173

Unidentified streltsy flag. According to the label sewn on the flag cloth, the banner was "taken in the Battle of Saladen near Birzein Lithuania on March 19, 1703." Dimensions: width 225 cm, height 210 cm, length of staff 360 cm. Note the decoration on the flag staff.

Armémuseum:AM.082166

Unidentified streltsy flag. According to the label sewn on the flag cloth,"the banner was taken in the Battle of Saladen near Birze in Lithuania on March 19, 1703."

Dimensions: width 250cm, height 215cm, staff's length 335cm.

Armémuseum:AM.082169

Unidentified streltsy flag. According to the museum, it is possible that this banner was taken in the Battle of Saločiai/Saladen outside Birzein 1703.

Dimensions: height 221cm, length of staff 335.5cm, width maximum 211cm.

Armémuseum: AM.082170

Unidentified streltsy flag. According to the label sewn on the flag's cloth, "the banner was taken in the Battle of Saladen near Birze in Lithuania on March 19, 1703." Dimensions: height 192 cm, width 222 cm, length of staff 242 cm.

Armémuseum: AM.082167

Unidentified streltsy flag. This banner has been identified as one of the four of the same kind included in Magnus Gabriel de la Gardie's list of trophies taken at Riga in 1656 *(RA Livonica II vol 77). "4 large Streltsy flags from Moscow and Lord Jacob's army."* According to a label on the pole, this banner was supposed to have been taken in Narva in 1700. This is contradicted by the fact that it was depicted earlier in Hoffman-Jonsson's *Depiction of the State's Trophy Collection* and that it is described in the 1685 armory listing and they are listed under the probably inaccurate heading "Polish soldier scenes."

Dimensions: width 310 cm, height 240 cm, staff's length 350 cm.

Armémuseum: AM.082196

In the museum's 1800 inventory this banner is listed under the heading: "*Taken at Janitzskå on February 2,1703*" and described as: "*1 piece of light blue gaiter, with blue and red crests in wavy lengths: across the cloth abroad red cross, with gilded edges, and a crown at each end, round a very gilded foliage, painted – broken.*"

A comparison with another flag in the museum shows that the same fabrics and decorative elements can be found on it as well as two additional colors. From this it can be assumed that this flag is a **company banner for the Smolensk Regiment under Colonel Vilim Iarlovitsch Leikin**. It was produced on December 17, 1692 (presumed date, provided that the flag is connected to a similarly constructed flag (ST 21:46) which bears that date).

Dimensions:width203cm,height187cm,staff'slength323cm.

Armémuseum:AM.082206

Unidentified streltsy flag. In the Swedish Army Museum's digital collection, this flag is simply listed as a "Russian streltsy regimental banner." According to the label on the pole, the flag was taken in Narva in 1700.

Dimensions:width217cm,height200cm,staff's length305cm.

Armémuseum:AM.082189

Unidentified streltsy flag. In the museum's 1800 inventory, this banner is listed under the Narva Trophies. However, according to the museum, "the information is uncertain. Possibly it may have been one of the banners described in the1703 inventory." "Flag is blue, white and red with a gold cross." Another possibility is that the flag may have been taken at the battle of Saladen on March 19,1703.It is made of damask,with dimensions: width 237.0cm and height 210.0cm.

Armémuseum:AM.082177

Unidentified streltsy flag. In older inventories it has been noted that this somewhat similar banner may possibly have been taken in the Battle of Saladen on March 19, 1703. In the list of captured trophies, this one is described as being made of red and gold damask.

Dimensions: width 217cm, height 220cm, staff's length 340cm.

Armémuseum:AM.082179

Unidentified streltsy flag. In the Swedish Army Museum's digital collection, this flag is simply listed as a "Russian streltsy regimental banner" with no further attribution to when or where it was captured.

Dimensions: width 222cm, height 205cm, staff's length 205cm.

Armémuseum:AM.082181

Publishedby:
OnMilitary Matters
31West BroadSt.
Hopewell,NJ08525USA 609-
466-2329
militarymatters@att.netwww.
onmilitarymatters.com

www.ingramcontent.com/pod-product-compliance
Lightning Source LLC
LaVergne TN
LVHW071633100826
845154LV00008BA/142
9780965328418